RAND NATIONAL DEFENSE RESEARCH INSTITUTE

Metrics to Compare Aircraft Operating and Support Costs in the Department of Defense

Michael Boito, Edward G. Keating, John Wallace, Bradley DeBlois, Ilana Blum

Prepared for the Office of the Secretary of Defense

For more information on this publication, visit www.rand.org/t/rr1178

Library of Congress Control Number: 2015959715
ISBN: 978-0-8330-9189-5

Published by the RAND Corporation, Santa Monica, Calif.

Preface

This report emanates from a RAND project titled "Developing a Consistent Definition of Cost per Flying Hour for Use Throughout the Department of Defense." The objective of this research project was to assist the Office of the Assistant Secretary of Defense for Logistics and Materiel Readiness (OASD [L&MR]) in developing a consistent definition of aircraft operating and support (O&S) cost per flying hour that can be used across different aircraft platforms in the Department of Defense (DoD). As the project evolved, the research team also examined the possibility of alternative metrics that high-level DoD decisionmakers can use to compare the O&S costs of different aircraft.

This research should be of interest to DoD personnel involved with weapon-system acquisition and O&S cost issues. It was sponsored by OASD (L&MR) and conducted within the Acquisition and Technology Policy Center of the RAND National Defense Research Institute, a federally funded research and development center sponsored by the Office of the Secretary of Defense, the Joint Staff, the Unified Combatant Commands, the Navy, the Marine Corps, the defense agencies, and the defense Intelligence Community.

For more information on the RAND Acquisition and Technology Policy Center, see http://www.rand.org/nsrd/ndri/centers/atp.html or contact the director (contact information is provided on the web page).

Contents

Figures

Tables

Summary

This report examines what definition of operating and support (O&S) cost per flying hour (CPFH) is best suited to compare the O&S costs of different aircraft, and how the costs and aircraft usage should be normalized when making comparisons. Such comparisons could inform choices by Department of Defense (DoD) decisionmakers regarding acquisition of new aircraft and retirement or retention of existing aircraft. The CPFH metric could also be used to monitor the progress of aircraft programs in meeting O&S affordability constraints.

CPFH is a well-known DoD cost metric. CPFH is calculated as an aircraft fleet's O&S costs divided by its flying hours.

CPFH is widely used by the military services for different purposes. One common usage is for the flying-hour programs (FHPs) used by the services to budget resources to achieve aircrew proficiency. The FHPs use a CPFH defined by DoD guidance that is intended to include only costs that vary with flying hours. FHP decisionmakers want to assess the budgetary impacts that incremental changes in flying-hour levels have on certain cost elements that vary directly with changes in flying hours (i.e., directly funded fuel, consumable materials and repair parts, depot-level reparables).

A related usage of CPFH is for flying-hour reimbursable billing rates, i.e., how much other DoD, other federal, other customers, and foreign military should be charged on a per-flying-hour basis. These rates build upon the FHP CPFH, adding in cost categories less directly impacted by flying hours such as depot-level maintenance costs.

Another, more challenging usage of CPFH is to compare the O&S costs of different aircraft programs. Typically, these comparisons are between a prospective new system and an antecedent system and are made between their average annual O&S costs.

A key difference between the CPFH used for FHP and reimbursable billing and the CPFH used to compare O&S costs of different aircraft programs is that cross-system O&S comparisons intentionally include some cost categories that are fixed, i.e., do not vary with flying hours. Used in other contexts, such as for decisions about the acquisition of new aircraft programs or the retention or retirement of existing aircraft force structure, decisionmakers likely care about a broader swath of O&S costs including those that vary with flying hours as well as those that do not.

When comparing O&S costs of different aircraft, we recommend that the O&S costs included in the comparison should be clearly defined. The DoD's Cost Assessment and Program Evaluation (CAPE) office defines a standard O&S cost-element structure that comprises six major elements of: (1) unit personnel, (2) unit operations, (3) maintenance, (4) sustaining support, (5) continuing system improvements, and (6) indirect support. In most instances, we recommend that all the direct elements of the standard O&S cost-element structure, or elements one through five, should be included when comparing O&S costs. A shorthand term for these costs is direct O&S costs. We recommend that indirect costs be excluded because, as their name implies, indirect costs are only indirectly affected by the choice, and because they are not captured consistently by the services' official O&S cost databases.

In addition to a consistent definition of O&S costs, we recommend a consistent measure of the number of flying hours per aircraft be used when comparing costs of different aircraft. Costs of fleets should be compared using stable annual flying-hour levels needed to achieve crew proficiency and exclude flying hours for contingency operations, because variations in flying hours per aircraft affect the calculations.

When comparing O&S costs of different aircraft, some basic data-normalization steps will be needed. Costs should be compared using constant dollars to normalize for the effects of inflation at different points in time. Costs should be compared at maturity, i.e., when the fleets are at their maximum size, because the ramp-up, steady-state, and ramp-down stages of the O&S phase tend to affect fleet cost.

Another normalization, which is difficult to perform especially when comparing new aircraft programs to their antecedent, is to account for differences in actual costs versus estimated costs. Actual flying hours and costs are typically constrained by available resources, while estimated flying hours and costs are typically based on the premise of full funding needed to achieve crew proficiency.

When the appropriate standardization steps mentioned above are taken, decisionmakers could use CPFH to compare O&S costs of different aircraft. Comparisons of CPFH are most appropriate when the intention is to compare costs that vary closely with flying hours, such as fuel, depot-level reparable, or perhaps engine-related costs. In most instances, when a broader collection of O&S costs is of interest, the O&S costs will include elements that are largely fixed or insensitive to changes in flying hours, such as unit-level personnel, sustaining support, or modifications. For this broader collection of O&S costs that includes costs that are largely fixed with respect to flying hours, CPFH changes inversely with flying hours, so that, for example, CPFH increases as flying hours and total fleet costs decrease. For this reason it is important that flying hours per aircraft be normalized as closely as possible and specified when comparing CPFH of different aircraft.

Alternative Affordability Metric

A metric that is an alternative to CPFH as a way to compare O&S costs of different aircraft is annual O&S cost per aircraft. Annual O&S cost per aircraft has the desirable characteristic of changing in the same direction that flying hours and total O&S costs for a fleet change. One subject-matter expert described efforts to reduce total program O&S costs, such as streamlining the flying-hour program and making more use of simulators rather than flying for training. These efforts reduce total program costs but increase CPFH, a counterintuitive result for an affordability metric.

In addition, a unique value of a military aircraft is its readiness and availability for tasking. Regardless of how much it is flown (assuming equipment and crew are mission capable), there is military value in force structure. A significant portion of the cost to achieve a force structure with available aircraft is largely fixed and insensitive to flying hours. Annual O&S cost per aircraft more intuitively expresses the cost of available aircraft.

When an annual O&S cost-per-aircraft metric is used, we recommend primary aircraft inventory (PAI) as the measure of the number of aircraft. PAI is the number of aircraft assigned to perform the mission and includes combat, combat support, training, and test aircraft. The number of PAI primarily determines the resources programmed for a fleet.

Acknowledgments

Molly Mertz of the Office of the Assistant Secretary of Defense for Logistics and Materiel Readiness (OASD[L&MR]) has been a dedicated and engaged action officer, and we are grateful for her support of the project. We also thank her colleagues Mark Gajda, James Kelly, and Larry Klapper.

We appreciate the time and insights we received from our subject-matter experts:

- Jennifer Bowles and her colleagues at the Air Force Cost Analysis Agency (AFCAA)
- Anthony Boyda, Breon Mitchell, and Rachel Shear of the United States Army
- David Cadman and his colleagues of Performance Assessments and Root Cause Analyses
- Tim Conley and his colleagues at the United States Navy Naval Air Systems Command
- Tim Echard of Lockheed Martin
- Ronald Fischer of the United States Air Force, Directorate of Strategic Planning
- Kimberly Fuller of the Joint Strike Fighter program office
- Thomas Henry of the Office of the Secretary of Defense, Cost Assessment and Program Evaluation (Office of the Secretary of Defense CAPE)
- Carlton Lavinder and his colleagues of the Naval Center for Cost Analysis

- Colonel Brian Stokes of the United States Army, Chief, Acquisition Support Program Analysis Division.

Tim Conley reviewed a draft of this report and provided substantive comments and insights that strengthened our understanding of the issues and improved the report.

At RAND, Cynthia Cook and Marc Robbins provided helpful program management for this project. Megan Bishop assisted with our literature review. Kathryn Connor and Jack Graser provided careful reviews and constructive feedback that improved the report. Linda Theung edited this report.

Of course, all remaining errors and interpretations are solely the authors' responsibility.

Abbreviations

AFCAA	Air Force Cost Analysis Agency
AFTOC	Air Force Total Ownership Cost
AT&L	Acquisition, Technology, and Logistics
CAPE	Cost Assessment and Program Evaluation
CMD	command
CPFH	cost per flying hour
DLR	depot-level reparable
DoD	Department of Defense
FH	flying hours
FHP	flying-hour program
FHPB	flying-hour program budgeting
FY	fiscal year
GAO	Government Accountability Office
ISR	intelligence, surveillance, and reconnaissance
L&MR	Logistics and Materiel Readiness
MD	mission design

O&S	operating and support
OSA	operational support airlift
OSD	Office of the Secretary of Defense
PAA	primary authorized aircraft
PAI	primary aircraft inventory
POL	petroleum, oil, and lubricants
SAR	Selected Acquisition Report
TAI	total active inventory
VAMOSC	Visibility and Management of Operating and Support Costs

CHAPTER ONE

Introduction

This report, and the project that generated it, started from a seemingly straightforward question: What are the most appropriate metrics that high-level Department of Defense (DoD) decisionmakers can use to compare the operating and support (O&S) costs of different aircraft? Such a comparison could inform acquisition choices (e.g., Is a prospective replacement aircraft worth purchasing?) or decisions about retaining existing fleets and shaping force structure (e.g., Are O&S cost trends for different aircraft evolving favorably or unfavorably?).

This seemingly straightforward question encounters several complications in real-world application. The complications include whether all elements of O&S costs should be included or just a subset of the total. In particular, should costs that are relatively fixed each year, such as personnel costs, be included? Or should only clearly variable costs, such as those for fuel and consumable and reparable parts, be included? How should costs be normalized when comparing costs for aircraft at different stages in the O&S phase, or with vastly different usage rates, or with different capabilities? In this report, we discuss these issues for metrics of aircraft O&S costs, as well as appropriate implementation of these metrics.

Cost per flying hour (CPFH) is a well-known DoD cost metric. As the name suggests, CPFH is calculated as an aircraft fleet's costs divided by its flying hours:

$$CPFH = \frac{Total\ O\&S\ Costs}{Total\ Flying\ Hours}\ .$$

There is, however, ambiguity as to which elements of O&S cost to include in the numerator as well as whether to use only peacetime flying hours in the denominator or whether to also include contingency or operational flying hours. Obviously, the more cost categories that are included in the numerator and the fewer flying hours that are included in the denominator, the greater the estimated CPFH will be.

The use of CPFH in some contexts is prescribed and defined in DoD policy.[1] Two such widespread uses of the term are for estimating budgets for the services' flying-hour programs (FHPs) and to determine hourly rates for aircraft that are used on a cost-reimbursable basis. Policy for usage of CPFH in the FHP requires that only certain costs that vary with flying hours (most obviously fuel, but also consumable parts and depot-level reparables [DLRs]) be included in the cost numerator. Excluded from the FHP or reimbursable calculations are costs for most unit-level personnel,[2] sustaining support, and modifications. As we discuss below, DoD leaders who are making decisions about acquisition or force-structure issues likely care about a broader swath of O&S costs than just the variable costs, i.e., the FHP-based calculation of CPFH may be inadequate.

A decisionmaker on acquisition and/or force-structure questions could use a different set of O&S cost elements and/or a different metric. Cost per aircraft—a fleet's O&S costs divided by the number of aircraft—is a possible alternative metric:

$$\textit{Cost per Aircraft} = \frac{\textit{Total O\&S Costs}}{\textit{Number of Aircraft}}\ .$$

While this metric has its own challenges, as we discuss below, it has advantages relative to CPFH as a cost metric. There are other pos-

[1] The exact citations in the DoD Financial Management Regulations are provided and summarized in Chapter Three.

[2] Reimbursable rates for non-DoD customers include hourly crew costs.

sible metrics, including cost per squadron, cost per fleet, and cost per capability, that could be useful for acquisition- and/or force structure–oriented decisionmakers.

The RAND research team used a variety of methodologies on this project. We reviewed DoD regulations and guidance, cited in Chapter Three. We examined the literature on CPFH. Perhaps most importantly, the research team conducted a series of subject-matter expert interviews. These subject-matter experts, who are listed in the Acknowledgments of this report and included employees of three military services, the Office of the Secretary of Defense (OSD), a program office, and a defense contractor, presented a variety of perspectives. While we draw on their views in this report, we do not attribute perspectives to named individuals.

The remainder of this report is structured as follows: Chapter Two presents background information on this project. Chapter Three summarizes ways the CPFH metric is used in DoD and also discusses data normalizations needed to make meaningful comparisons of the O&S costs of two or more aircraft. Chapter Four contains a discussion of alternative O&S cost metrics. Chapter Five provides our recommendations for O&S cost metrics to be used when comparing aircraft. Appendix A presents an example of how CPFH and cost per aircraft change when flying hours are reduced. Appendix B presents the results of a statistical analysis of Air Force O&S cost elements to test the statistical correlation between the cost of each element and flying hours and the number of aircraft.

CHAPTER TWO

Background

This six-month project that commenced in the late spring of 2014 was titled "Developing a Consistent Definition of Cost per Flying Hour for Use Throughout the Department of Defense." The objective of this research project was to assist the Office of the Assistant Secretary of Defense for Logistics and Materiel Readiness (OASD[L&MR]) in developing a consistent definition of aircraft CPFH that can be used across different aircraft platforms in DoD.

The project had four tasks:

1. Review and document all uses of the term *CPFH* within the department and across the services.
2. Review and document how program offices have historically estimated and compared CPFH of new programs to legacy programs.
3. Recommend a consistent definition of CPFH for DoD aviation platforms and if another type of metric would be more useful in making O&S cost comparisons.
4. Recommend methodology for comparing CPFH of a new weapon system with a legacy system (e.g., F-35 compared to F-16).

Subject-matter experts told us that the Under Secretary of Defense for Acquisition, Technology, and Logistics (USD[AT&L]) wanted one or a few metrics to compare aircraft O&S costs. CPFH has been used for this purpose, but it was sometimes found to have been estimated differently (e.g., inclusion of different cost categories in the numerator) across aircraft systems. In addition, because of changes in total flying hours, CPFH can decrease when total O&S costs increase and vice versa, as explained later, which makes it counterintuitive as an indication of affordability.

Before proceeding to the detailed discussion of O&S costs in following chapters, it is helpful to review the universe of O&S costs in DoD and the DoD standard O&S cost-element structure. These are defined in the OSD Office of Cost Assessment and Program Evaluation's (CAPE's) *Operating and Support Cost-Estimating Guide* (OSD, 2014). The cost-element structure has six major elements, with various levels of indenture below the six major elements. The cost-element structure at one level of indenture below the six major elements is shown in Table 2.1 and is taken directly from the cost-estimating guide. The second column of Table 2.1 shows our assessment of the relationship of each element of cost for aircraft systems to flying hours.

Although most of the elements are largely self-explanatory, a couple of elements merit some explanation. Element 2.1—operating material—overwhelmingly reflects fuel costs. Hardware modifications—Element 5.1—"excludes modifications undertaken to provide additional operational capability not called for in the original system design or performance specifications" (OSD, 2014, p. 6-15).

The first five major elements of the cost-element structure are directly related to the weapon system, and the sixth major element contains indirect costs. "Indirect support costs are those installation and personnel support costs that cannot be identified directly (in the budget or FYDP[1]) to the units and personnel that operate and support the system being analyzed, but nevertheless can be logically attributed to the system and its associated manpower" (OSD, 2014, p. 6-16). "Indirect support costs are more relevant for situations in which total DoD manpower would change significantly or when installations are affected (i.e., expanded, contracted, opened, or closed)" (OSD, 2014, p. 6-16). The CAPE O&S guide indicates that indirect costs should only be included in cost estimates when those costs would likely change due to the action being analyzed. Some, but not all, indirect costs are

---

[1] The Future Years Defense Program is the five-year program and financial plan for DoD.

included in the Air Force's official O&S cost-reporting system, Air Force Total Ownership Cost (AFTOC). Indirect costs are not included as part of the standard reports for weapon system costs in the Navy's official cost-reporting system, Visibility and Management of Operating and Support Costs (VAMOSC).[2]

Table 2.1
DoD Standard Cost-Element Structure and Relationship of Costs to Flying Hours

Category	RAND-Assessed Relationship to Flying Hours
1.0 Unit-Level Manpower	**Fixed**
1.1 Operations	Fixed
1.2 Unit-level maintenance	Fixed
1.3 Other unit level	Fixed
2.0 Unit Operations	
2.1 Operating material	Variable
2.2 Support services	Fixed
2.3 Temporary duty	Fixed
2.4 Transportation	Fixed
3.0 Maintenance	
3.1 Consumable materials and repair parts	Variable
3.2 Depot-level reparables	Variable
3.3 Intermediate maintenance	Variable
3.4 Depot maintenance	Semi-variable[a]
3.5 Other maintenance	Undefined/Unknown

[2] The Air Force's official O&S cost reporting system is the AFTOC decision support system. The Navy's official O&S cost reporting system is the VAMOSC management information system. Both are available online with restricted access.

Table 2.1—Continued

Category	Rand-Assessed Relationship to Flying Hours
4.0 Sustaining Support	Fixed
4.1 System-specific training	Fixed
4.2 Support equipment replacement and repair	Fixed
4.3 Sustaining/systems engineering	Fixed
4.4 Program management	Fixed
4.5 Information systems	Fixed
4.6 Data and technical publications	Fixed
4.7 Simulator operations and repair	Fixed
4.8 Other sustaining support	Fixed
5.0 Continuing System Improvements	Fixed
5.1 Hardware modifications	Fixed
5.2 Software maintenance	Fixed
6.0 Indirect Support	Fixed
6.1 Installation support	Fixed
6.2 Personnel support	Fixed
6.3 General training and education	Fixed

[a] Depot maintenance costs for engine overhauls have a time-lagged and indirect relationship to flying hours. If engine modules are treated as a DLR component, those overhaul costs are included in Element 3.2.

Our characterization of fixed and variable costs requires explanation. Most of the elements shown are affected at least somewhat by both flying hours and the number of aircraft. Our characterization of elements as either fixed or variable indicates whether that element is predominantly affected by flying hours. Fixed costs are largely stable over a defined, forecasted range of activity. If that level of activity is increased or decreased significantly, especially over a foreseeable amount of time, fixed costs would no longer be fixed. For example, if flying hours are doubled over a sustained period, it is highly probable that numbers of maintenance personnel and pilots would have to be increased.

Similarly, costs we categorize as variable can include some fixed portion that is unaffected by flying hours. For example, 2015 Air Force Working Capital Fund budget estimates suggest that roughly 25 to 30 percent of consumable and reparable costs are overhead costs of operating the supply chain for those parts and are mostly fixed (see U.S. Air Force, 2014).

To get an idea of the proportion of fixed and variable aircraft O&S costs, we examined the Air Force's total direct aircraft O&S costs in fiscal years (FY) 2010 through 2013 as reported in AFTOC and the Navy's total direct aircraft O&S costs over the same period in VAMOSC in the standard O&S cost-element structure. These costs exclude indirect costs. We used FY 2014 constant dollars. We observed that the proportions of cost by element did not change much from year to year. We totaled the constant dollars by element over the four years and calculated the percentage of the total direct cost for each element. If we aggregate Table 2.1's categories to the one-digit level (e.g., 1.0, 2.0, 3.0), we find that one-digit categories with some variable component (2.0 and 3.0) accounted for 59 percent of total direct Air Force aircraft O&S costs and 57 percent of total direct Navy aircraft O&S costs, while the remaining costs are fixed. The results of this aggregation are shown in Table 2.2.

Our characterization of costs can be tested statistically using historical cost data, although the statistical results must be interpreted with caution. For some elements, statistical analysis will not show a strong correlation between costs of the element and flying hours because costs for the element are affected by other factors. For example, aircraft overhauls are generally scheduled on a chronological basis that is influenced by the design of the aircraft as well as its average flying hours per year. Similarly, engine overhauls are a function of the reliability of the engine and cumulative flying hours or cumulative cycles, which are a function of flying hours. Thus a correlation analysis between annual flying hours and airframe or engine overhauls is not likely to show a significant relationship because the overhauls result in part from cumulative rather than annual flying hours, as well as other factors.

Table 2.2
Air Force and Navy Aircraft O&S Cost Percentages by Element in FY 2010–2013

O&S Cost Element	Percentage of Direct Air Force O&S Costs	Percentage of Direct Navy Aircraft O&S Costs*	Aggregation of Table 2.1 Assessment of Relationship to Flying Hours
1.0 Unit personnel	30	27	Fixed
2.0 Unit operations	26	17	Semi-variable
3.0 Maintenance	33	40	Semi-variable
4.0 Sustaining support	2	2	Fixed
5.0 Continuing system improvements	9	13	Fixed

* This column does not equal 100 percent due to rounding.
SOURCES: AFTOC and VAMOSC.
NOTE: AFTOC is available with restricted access on the Air Force Portal at https://aftoc.hill.af.mil. Data were accessed on October 10, 2014. VAMOSC is available with restricted access at https://www.vamosc.navy.mil. Data were accessed January 8, 2015.

A statistical analysis of data is described in Appendix B of this report. The statistical analysis generally supports RAND's assessment of fixed and variable categories. The high degree of correlation between flying hours and total active inventory (TAI), however, makes interpretation of the statistical results difficult at best and inconclusive at worst.

Table 2.2 illustrates that a significant amount, roughly two-fifths, of O&S costs are in the elements of 1.0, unit personnel; 4.0, sustaining support; and 5.0, continuing system improvements. As indicated in the table, the costs in these elements are largely fixed and thus do not change when flying hours change. A CPFH metric that includes these fixed elements of direct O&S costs will vary inversely with flying hours; that is, the CPFH metric will increase as flying hours decrease and vice versa. Appendix A provides an example of how CPFH increases when flying hours are reduced.

Conversely, if CPFH includes only variable costs (which is what is intended in the FHP budgeting usage of the concept), CPFH will be invariant to moderate changes in the level of flying hours.

A related difficulty with the CPFH metric is that the denominator of flying hours for a given fleet tends to change over time due to contingency flying and budget availability. Flying hours are a policy lever to reduce total O&S costs. Flying hours are therefore unstable over time and make the CPFH metric volatile.

In the next chapter, we discuss the widely used CPFH metric.

CHAPTER THREE

Applications of Cost per Flying Hour in DoD

The term *cost per flying hour* (CPFH) has been used in different contexts in DoD. In this chapter, we discuss its usage in budgeting for the services' flying-hour programs, for reimbursable rates for customers that use DoD aircraft, in responding to requests for information from outside DoD, and for cross-system comparisons. Later in the chapter, we discuss appropriate steps to normalize CPFH for cross-system comparisons.

Different Contexts for Use of CPFH

Flying-Hour Program Budgeting

One usage of CPFH has been in FHP budgeting (FHPB). The typical question here is if one wishes to adjust flying hours incrementally (up or down), how much must the FHP budget change? Or, to accommodate a given budget change, how much must flying hours change? Instructions for preparing a Flying Hours Program Exhibit OP-20 are presented in "Operation and Maintenance Appropriations" (DoD, 2010). The CPFH used for FHPB is intended to capture only elements of cost that are directly variable with flying hours. These elements are fuel, consumables, and DLRs.

$$CPFH_{FHPB} = \frac{Fuel + Consumables + DLRs}{Flying\ Hours}.$$

The guidance is applicable to all the services.

Reimbursement Rates

A related usage of CPFH is for flying-hour reimbursable billing rates, i.e., how much other DoD, other federal, other customers, and foreign military should be charged on a per-flight-hour basis.[1] Instructions for collections of reimbursements for use of DoD-owned fixed-wing aircraft are presented in "Collections for Reimbursements of DoD-Owned Aircraft (Fixed Wing)" (DoD, 2011).[2]

The prescribed rate for DoD customers is to include the cost categories typically found in an FHP CPFH, i.e., fuel, consumables, and DLRs, but it also includes depot-maintenance costs and, if applicable, variable contractor logistics support (CLS) costs.

$$CPFH_{Reimb} = CPFH_{FHPB} + \frac{Depot\ Maintenance + Variable\ CLS^{3}}{Flying Hours}.$$

The DoD customer-reimbursement rate should be greater than or equal to the FHP CPFH. The other federal agency rate and the foreign military sales rate cover the DoD rate's cost categories, plus also include an allocation of crew salary costs, not considered in the DoD rate or in FHP CPFH.

[1] Such cost factors have been used to estimate the search costs associated with the missing Malaysia Airlines Flight 370. *World Maritime News* (2014), for instance, quotes Pentagon spokesman Army Col. Steve Warren as saying a P-8 Poseidon aircraft costs $4,200 per flight hour.

[2] This rate does not, however, apply to flights such as sustainment air missions, for which there is a commercial analog. When there is a direct commercial alternative, customers are charged rates benchmarked against that commercial alternative. If customer-generated rates are insufficient to cover DoD costs, the Airlift Readiness Account, funded by direct Air Force appropriation, covers the shortfall (see DoD, 2009). Of course, many types of DoD aircraft lack commercial analogs, in which case CPFH flying-hour reimbursable billing rates apply.

[3] CLS can be fixed or variable in nature, but only variable CLS costs are intended to be captured here.

$$CPFH_{Other\ Federal\ Agency} = CPFH_{Reimb} + \frac{Allocation\ of\ Crew\ Salary\ Costs}{Flying\ Hours}.$$

The public rate includes all of the aforementioned cost categories plus an allocation for asset utilization (depreciation) and unfunded civilian retirement costs.[4]

$$CPFH_{Public} = CPFH_{Other\ Federal\ Agency} + \frac{Depreciation + Allocation\ of\ Unfunded\ Retirement\ Costs}{Flying\ Hours}.$$

DoD Instruction 4500.43 Change 1 provides instructions for operational support airlift (OSA) aircraft, including specifying the costs per flying hour to be considered for cost-effectiveness comparisons to commercial air travel. "The aircraft operating cost per flying hour for OSA aircraft missions shall include petroleum, oil, and lubricant costs [POL]; unit intermediate and depot-level maintenance, including civilian and contract maintenance labor; spares; and crew per diem costs" (Under Secretary of Defense for Acquisition, Technology, and Logistics, 2013).

$$CPFH_{OSA} = \frac{POL + Maintenance + Spares + Crew\ per\ diem}{Flying\ Hours}.$$

Responding to Congressional and Media Requests for Information

Many requests for information from Congress and media on the O&S cost of an aircraft are couched in terms of CPFH. Typical questions are: How does CPFH of one aircraft compare to another or what does it cost to fly a particular aircraft? The action officer or individual given the responsibility of responding to the inquiry would typically choose the definition of CPFH appropriate to each specific question. In some situations, $CPFH_{Reimb}$ is provided; in others, it may be $CPFH_{FHPB}$.

---

[4] FY 2014 hourly reimbursable rates by airframe and customer (Office of the Under Secretary of Defense [Comptroller], 2013).

An example is the flurry of requests for information to DoD after an aircraft used as Air Force One and two F-16 fighters flew by the Statue of Liberty at low altitude in 2009. The flight was made to take a publicity photograph to update the file photo of Air Force One. Cost estimates for the three-hour mission involving three aircraft were reported as $328,000 (CNN, 2009). The letter from then–Secretary of Defense Robert Gates to Senator John McCain regarding the incident and its cost (Gates, 2009, p. 2) explained that the cost included:

> primarily fuel, depot level repairables, and consumables for the F-16 and fuel only for the VC-25 (since it is maintained through a contractor logistics support contract). The standard methodology prescribed in the OSD Financial Management Regulation includes not only these reimbursables but also annualized average maintenance costs allocated on a per flying hour basis. This includes depot level maintenance, engine overhaul, and all contractor logistics support costs characterized as variable.

This response corresponds to $CPFH_{Reimb}$.

Monitoring One Aircraft System's Costs over Time

CPFH is used to monitor an aircraft system's cost trends over time. A variation of this idea is that certain subsets of aircraft CPFH that are highly variable with flying hours, such as DLRs or engine O&S costs, are monitored for cost trends. Identification of unfavorable trends in CPFH can motivate root-cause analyses and corrective actions, such as development of reliability improvements.

While CPFH is well suited for monitoring an aircraft system's cost trends over time, it is subject to distortions over time discussed previously; that is, when costs that are fixed or largely insensitive to flying hours are included, CPFH varies inversely with flying hours per year.

Cross-System O&S Cost Comparisons

Another and more challenging usage of CPFH is to compare weapon systems' O&S costs. Typically, these comparisons are between a prospective new system and an antecedent system and are made between their average annual O&S costs.

A key difference between CPFH used in the FHP or for reimbursement rates and CPFH used for cross-system O&S cost comparisons is that cross-system O&S comparisons intentionally include some cost categories that are fixed and do not vary with flying hours. Fixed costs, such as unit-level mission personnel, sustaining engineering, software maintenance, and modification investments, can vary considerably across weapon systems and can be thus of interest to a cross-system O&S cost comparison.[5]

$$CPFH_{Cross\text{-}System} = CPFH_{Reimb} + \frac{Fixed\ Costs}{Flying\ Hours}.$$

We reiterate that fixed costs are those that are invariant with respect to flying hours. They might vary considerably, however, across two different weapon systems, so it is entirely appropriate to consider them in a cross-system comparison.

In our subject-matter experts interviews, we learned that analysts have used CPFH in cross–weapon system cost comparisons. Comparisons we heard of included F-35 versus legacy fighter aircraft,[6] Global Hawk versus U-2, and CSAR-X versus UH-60.[7] These comparisons all used some implementation of the $CPFH_{Cross\text{-}System}$ metric.

By contrast, the United States Government Accountability Office (GAO) (2009) used a "cost per flight hour" metric defined by adding the total cost of fuel, flight equipment, consumables, and reparables,

---

[5] For example, Anneker, Germony, and Pardoe (2014) enumerate a number of flying-hour invariant cost categories that they argue should be included in an unmanned aerial system metric they term *Burdened Cost per Station Hour.*

[6] Note that Air Force aircraft were compared to Air Force aircraft and Navy/Marine aircraft to Navy/Marine aircraft. Cross-service comparisons of systems may require additional normalizations to adjust for different accounting and organizational practices of each service (e.g., differences in what each service considers direct versus indirect).

[7] For each of these pairs, the first system is the new system and the second system is the antecedent system.

then dividing by flight hours flown when comparing MV-22 CPFH to its antecedent, the CH-46E.[8] This measure is the FHPB cost per flying hour, $CPFH_{FHPB}$. We do not know whether the GAO's results would have been different had they instead used the $CPFH_{Cross\text{-}System}$ metric.

The term *CPFH* has been used by different analysts to mean different things. Figure 3.1 illustrates this issue. The Air Force Cost Analysis Agency (AFCAA) created Figure 3.1 as part of a discussion of the many ways F-35 CPFH could be defined, although the phenomenon of different definitions of the term *CPFH* is not unique to the F-35. Not all of Figure 3.1's categorizations directly map onto the CPFH definitions we have used heretofore, though the *Reimbursable CPFH* box corresponds to $CPFH_{Reimb}$. In that the *Reimbursable CPFH* box includes depot maintenance costs, none of Figure 3.1's categorizations is as narrow as $CPFH_{FHPB}$.

[8] U.S. Government Accountability Office correctly notes that "costs per flight hour for various aircraft should be considered in the context of their capabilities, missions flown, and actual usage" (GAO, 2009, p. 4, footnote 9).

Figure 3.1
Different Cost Elements Used in CPFH Comparisons

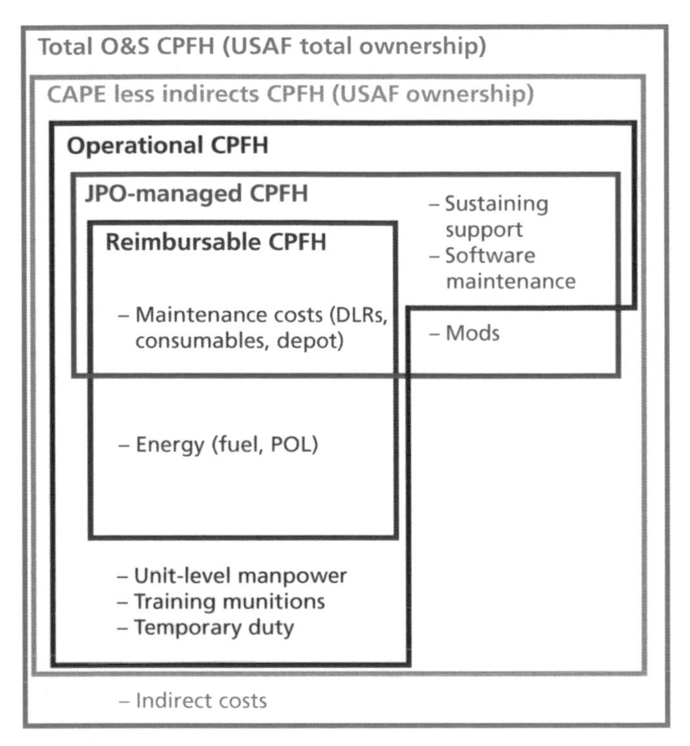

SOURCE: Air Force Cost Analysis Agency.
RAND *RR1178-3.1*

One key point regarding Figure 3.1 is that there are different versions of CPFH in use with different definitions of what is and is not included in the cost numerator. It is clearly important to use the same definition for an apples-to-apples cross–weapon system comparison so that the same kinds of costs are included for the systems being compared.

A second key point regarding Figure 3.1 is that the definitions and terminologies in it were devised by DoD cost analysts from multiple organizations trying to define a common terminology for referring to F-35 CPFH. At least among O&S cost estimators, there is widespread understanding of the need for a consistent definition of CPFH.

Our interviews with subject-matter experts bore this out. The issue, however, is not as well understood outside the O&S cost-estimating community.

Appropriate Normalization of CPFH

In this section, we summarize CPFH comparisons of the F-35 and F-16 and of the U-2 and RQ-4 (Global Hawk). We then generalize the discussion to normalizations that are advisable whenever O&S costs are compared, irrespective of whether CPFH or a different O&S comparison metric is used.

Here is an enumeration of the adjustments Air Force Coast Analysis Agency (AFCAA) personnel made to compare F-35A to F-16C/D CPFH for use in the F-35 selected acquisition reports (SAR):

- Normalized flying hours and costs to the same flying hour/primary authorized aircraft (PAA) rate
- Normalized fuel costs
- Normalized TAI to PAA ratio
- Used the same inflation indexes
- Normalized F-16C/D mission personnel costs to reflect authorized positions rather than the cost of assigned personnel reported in AFTOC
- Normalized budget-constrained expenditure data from AFTOC to reflect requirements
- Added weapon-system costs not found in AFTOC for the F-16, e.g., Low Altitude Navigation and Targeting Infrared for Night pods. These additions increased the F-16C/D CPFH by 4 percent.

The normalizations increased the F-16C/D CPFH above the raw costs reported in AFTOC by a few thousand dollars per flying hour. The normalized CPFH in budget year 2012 dollars reported in the 2013 SAR were $32,554 for the F-35A, and $25,541 for the F-16C/D.

In its budget submission for FY 2013, the Air Force proposed divesting itself of 18 RQ-4 (Block 30) Global Hawk aircraft and retaining the U-2. The Air Force's rationale for the proposal was a reduction in high-altitude intelligence, surveillance, and reconnaissance (ISR) combat air patrol requirements, the need for upgrades to the RQ-4 to meet current U-2 sensor capability, and higher O&S costs of the RQ-4 compared to the U-2. In response, congressional defense committees directed the Secretary of the Air Force to provide a report with detailed costs pertinent to the decision. The Air Force report illustrated the effect of normalizations for mission, flying hours, and capability.

The report states that costs per flying hour in FY 2012 for Global Hawk and U-2 were roughly equal (U.S. Air Force, 2013). The comparison assumed a high-altitude ISR mission from the same base at various ranges to the target area. As the distance and therefore transit time to the target area increases in the scenarios, the Global Hawk's greater endurance allows it to spend more time on station per sortie, fly fewer sorties, and cost less per mission as the range of each scenario increases. For the mission with the longest range to the target, the U-2 cost was 46 percent more than that of the Global Hawk. Thus, although the two aircraft have a roughly equal CPFH, normalizing for mission resulted in a lower cost per mission for Global Hawk, as the distance to the target increased. This difference, while not reflected in CPFH, would be reflected in a cost-per-aircraft measure.

The report also considered the costs of normalizing capability so that Global Hawk Block 30 would be retrofit with sensor capability at least as good as the U-2. The development effort was estimated to take several years and cost several hundreds of millions of dollars, in addition to the cost of retrofitting. The OSD O&S cost-estimating guide (2014) "excludes modifications undertaken to provide additional operational capability not called for in the original system design or performance specifications" from O&S costs and treats them as modernization costs instead. The Air Force report treated the costs of the capability upgrade consistent with this policy and did not include them in its CPFH, but estimated the costs of the capability increase and reported them as relevant to the decision to retire the Global Hawk Block 30. The Air Force concluded that it would be less expensive to retire the Global Hawk Block 30 and buy the Navy's MQ-4C Triton than to upgrade the Global Hawk.

Generalized Discussion of Normalizations

Whenever O&S costs are compared, the elements of O&S cost included should be the same for the aircraft being compared. Indirect costs should usually be excluded from the cost tabulation. The official O&S cost databases used by the services either do not report indirect costs at all or do not report the costs as prescribed by the OSD CAPE's *Operating and Support Cost-Estimating Guide*, making it difficult to obtain these costs. In any case, we believe that it is not desirable to include indirect costs. Decisionmakers are usually concerned with costs directly associated with the choice or decision. As their name implies, indirect costs are more loosely associated with a weapon system.

Costs should be compared in constant dollars using the same inflation indexes for the systems being compared. We acknowledge that while this is standard advice for comparing costs over different time periods, even when followed it is difficult to achieve the desired intent, especially when estimated future costs are involved. Some elements of O&S costs are volatile and impossible to predict, such as fuel prices. Costs for other elements are heavily affected by peculiar sectors of the economy, such as DoD industrial activities, the costs of which can be volatile, are impossible to predict, and can be difficult to normalize even retroactively. In general, comparisons of legacy to prospective fleets require use of future escalation factors that are estimates and therefore inherently uncertain.

Costs should be compared at maturity, i.e., when the fleets are at their maximum size.[9] Neither fleet should be ramping up nor ramping down at the point of comparison.

Costs should be analyzed over a multiyear time frame (if there are significant fluctuations in yearly costs around the point of comparison) so as to avoid disproportionate influence of one-year idiosyncrasies.

---

[9] This is equivalent to what Office of the Secretary of Defense terms *steady-state operations*: "The steady-state period begins when all systems are delivered, and ends when the first system retirements begin" (Office of the Secretary of Defense, 2014, p. 5-3). We note that this steady state may never obtain for the F-35 due to its planned 30 years of production, which could reasonably result in retirement of the earliest-produced aircraft before the last unit is delivered.

Costs should be compared using stable peacetime flying-hour levels, ideally at usage rates that are similar for the aircraft being compared. The flying hours per aircraft needed to achieve crew proficiency is a reasonable standard. Additional contingency flying hours almost certainly drive CPFH downward and cost per aircraft upward. The additional hours flown during contingencies can affect cost metrics in two ways. First and already noted, the inclusion of fixed costs in a CPFH metric results in a decrease in CPFH when flying hours increase, while total O&S costs and unitized costs increase. Second, contingency flying hours can differ from peacetime flying hours, e.g., with longer sortie durations and different mission profiles, which in turn affect CPFH (see, for instance, Wallace, Houser, and Lee, 2000).

Costs should be normalized to reflect authorized personnel levels rather than the costs of assigned personnel. Often units are manned at levels below authorizations, although during periods of abnormally high activity, such as the contingency operations in Iraq and Afghanistan, manning may be increased for affected units. Similarly, other historical costs affected by budget constraints should be normalized when compared to estimates based on unconstrained requirements. In cases in which differences in geographical location affect costs, these differences should be normalized.

Once these normalizations are performed, it is important to realize that normalized costs may differ significantly from actual costs. The normalizations are done for analytical purposes to allow valid comparisons across aircraft systems or over time for a given aircraft, and resulting normalized costs will be inappropriate for budgeting for non-normalized, real-world conditions.

When appropriate standardization and normalization steps are taken, CPFH can be used to compare aircraft O&S costs or to monitor trends in aircraft O&S costs over time. A CPFH metric is particularly appropriate when only the costs that are clearly variable with flying hours are of interest. If a CPFH metric is to be used to compare aircraft system costs, we recommend a CPFH that includes Elements 1-5 in the DoD O&S cost-element structure, known as *CAPE Less Indirects* in the parlance used in Figure 3.1. The *CAPE Less Indirects* CPFH excludes indirect costs that are not captured consistently in the services' official O&S cost databases and that are not directly influenced by the choice of aircraft system. We refer to these as direct O&S costs.

$$CPFH_{Cross\text{-}System} = \frac{Direct\ O\&S\ Costs}{Flying\ Hours}.$$

Next we discuss alternative O&S cost metrics. Note, however, that the normalization steps discussed in this chapter generally apply to CPFH and alternative O&S cost metrics as well.

CHAPTER FOUR

Alternative Metrics for Aircraft Operating and Support Costs

We asked about the use of CPFH as a measure of affordability in our interviews with subject-matter experts in cost analysis and in long-range planning and affordability assessments. Several of the subject-matter experts thought that CPFH is unsuitable as a measure of affordability. One subject-matter expert described efforts to reduce total program O&S costs, such as streamlining the flying-hour program and making more use of simulators rather than flying for training. These efforts reduce total program costs but increase cost per flying hour, a counter-intuitive result for an affordability metric.

The experts we interviewed in DoD voiced the same concern as the GAO in its report on F-35 O&S costs (2014) regarding the use of CPFH as an affordability metric. The GAO noted the CPFH targets for each F-35 variant, but argued that achieving CPFH targets would not necessarily imply that the program is affordable. The GAO used total estimated O&S costs constrained by service budgets as its primary metric for affordability. DoD experts generally concurred, with one planner saying succinctly that the only useful way of approaching affordability is to look at the enterprise force structure and programmed resources and consider resources for the weapon system in that total context.

On the basis of these concerns with CPFH, we explored alternative metrics to use when comparing O&S costs for aircraft systems. Cost per aircraft, cost per squadron (or unit), cost per fleet, and cost per capability are all possible alternative metrics. While these metrics are not perfect, they address, at least in part, concerns about the sensitivity of CPFH to flying-hour levels. Because CPFH has flying hours in the denominator of the ratio to costs, CPFH increases when flying hours (and total O&S costs) decrease. This phenomenon is illustrated in Figure 4.1, showing a near mirror-image relationship between KC-135 flying hours per aircraft and constant-dollar CPFH.

Figure 4.1
The Mirror-Image Relationship Between KC-135 Cost per Flying Hour and Flying Hours per Aircraft

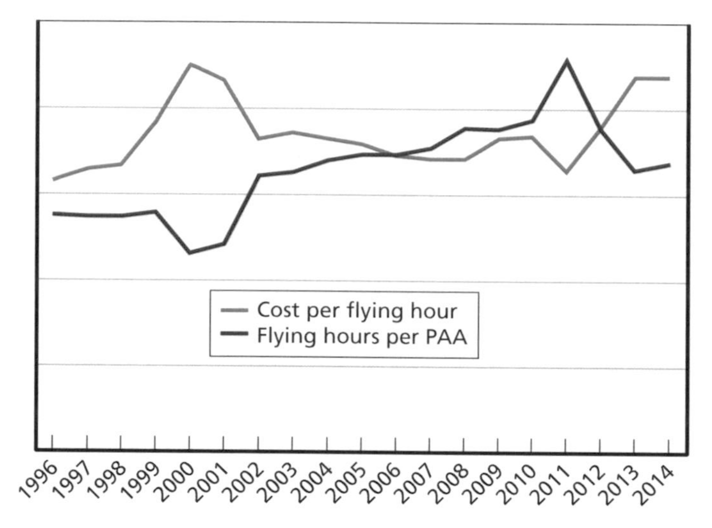

SOURCE: AFTOC.

When CPFH is used as an indication of efficiency or affordability, it is counterintuitive that the metric decreases when flying hours increase, potentially misleading decisionmakers who are not aware of this feature of the metric. But such behavior is to be expected when fixed costs are included in *Total O&S Costs* as in the previous chapter's $CPFH_{Cross\text{-}System}$ formula.

Cost per Aircraft

Several subject-matter experts spoke highly of cost per aircraft as a metric for comparing O&S costs across aircraft systems. For this metric, the number of total aircraft, not flying hours, serves as the denominator. As noted in Chapter One,

$$\textit{Cost per Aircraft} = \frac{\textit{Total O\&S Costs}}{\textit{Number of Aircraft}}.$$

Chapter Three discussed different views as to what should be included in the *Total O&S Costs* numerator. An additional ambiguity is what to use as the denominator. There are different possible definitions of total aircraft, including PAI and TAI. We use the terminology provided in the Chairman of the Joint Chiefs of Staff Instruction (2013), applicable to all military services. PAI is the number of aircraft assigned to perform the unit's mission and includes combat, combat support, training, and test aircraft. PAI excludes attrition reserve and backup inventory, which are included in TAI.[1] PAI is more stable over time than TAI because TAI includes attrition reserve aircraft, which decrease over time. Because attrition reserve aircraft, as well as the inventory of backup aircraft, are not used in the programming and budgeting processes to estimate required resources, PAI is a more appropriate denominator than TAI.

We therefore propose use of a metric we term

$$\textit{Cost per Aircraft (PAI)} = \frac{\textit{Total O\&S Costs}}{\textit{PAI}}.$$

[1] Attrition aircraft are procured specifically to replace the anticipated losses due to peacetime or wartime attrition. Backup aircraft are those intended to be in a maintenance pipeline (e.g., depot maintenance) and not available for mission operations.

A desirable characteristic of the cost-per-aircraft metric is that it increases or decreases in the same direction as changes in the number of flying hours. Appendix A illustrates the change in the cost-per-aircraft metric when flying hours are reduced.

With PAI, the cost numerator increases less than proportionally with an increase in flying hours, but the number-of-aircraft denominator is unchanged. Therefore, cost per aircraft increases when flying hours and total O&S costs increase. The result is intuitive for a cost metric.

Two other characteristics contribute to the desirability of the cost per aircraft (PAI) metric. First, while a majority of O&S cost is variable with flying hours, most O&S cost elements and costs, including unit-level consumption, are also variable with changes in PAI. Thus the cost per aircraft (PAI) metric will change less if there is a change in the number of PAI. Second, while flying hours are a significant policy lever for controlling costs and are easy to change, the number of PAI is less likely to change unless there is a restructuring of the acquisition program or existing force structure. Thus the cost per aircraft (PAI) is inherently more stable or likely to change less than CPFH, and changes in the metric move in the same direction as the increases or decreases in the O&S costs of the program.

Robbert (2013) presents a contrast between cost per aircraft and CPFH and their suitability as metrics in measuring the cost of meeting key demands. One important strategic demand is the ability to provide a large fleet of aircraft to meet surge requirements in a contingency. In meeting this strategic demand, cost per aircraft is a suitable metric. An important operational demand is the ability to provide operational flying hours. In meeting this operational demand, cost per flying hour is a suitable metric. Robbert finds that Air Force reserve-component units provide mission-ready aircraft at lower cost per aircraft than active units. In contrast, active units have often met operational demands at lower CPFH. This difference is driven by the fact that active units generally fly their assigned aircraft more hours per month.

For the purposes of this report, one takeaway is that each metric is suitable for measuring a different purpose, with cost per aircraft more suitable for measuring the cost of having aircraft available to fly, and

CPFH more suitable for measuring the cost to provide operational flying hours. The strategic value of maintaining readiness for wartime is unique to DoD. An advantage of comparing DoD aircraft O&S costs per aircraft is that it is a more suitable metric to assess the annual cost to maintain readiness.

Cost per Squadron or Unit

Cost per squadron is intended to provide a measure of cost per combat capability for aircraft with the same mission, assuming that squadrons are sized to achieve equivalent capability.

$$\textit{Cost per Squadron} = \frac{\textit{Total O\&S Costs}}{\textit{Number of Squadrons}}.$$

Uncertainty early in an acquisition program, however, about the number, size, and location of squadrons detracts from this metric's utility. In addition, squadron sizes often differ for a given fleet so that it is not obvious which squadron size to choose as representative of that fleet.

Cost per Fleet

Cost per fleet is a useful metric if one is concerned about overall affordability in the budgetary process.

$$\textit{Cost per Fleet} = \textit{Total O\&S Costs associated with weapon system}$$

In GAO (2014), the GAO uses total estimated O&S costs as its primary metric in expressing concerns about F-35 sustainment costs. The report notes CPFH targets for each F-35 variant, but the GAO is concerned that achieving these CPFH targets would not necessarily imply that the program is affordable. The GAO urged that total fleet costs be assessed against service budget constraints. Such affordability assessments are done by service-planning staff.

Figure 4.2 compares KC-135 constant-dollar CPFH, cost per aircraft (PAI), and cost per fleet, normalizing each series' FY 1996 value to 1.0. (With this normalization, for each metric, Figure 4.2 displays $\frac{\textit{Year N value of metric}}{\textit{1996 value of metric}}$). Since 1996, cost per aircraft has increased considerably, but fleet costs have not increased commensurably because the Air Force retired a number of KC-135Es.

Cost per fleet could be normalized by comparing two ratios. The first ratio is the cost of a given fleet to the cost of all of a service's fleets.

$$\frac{\textit{Total O\&S Costs of Fleet of Interest}}{\textit{Total O\&S Costs of All Air Force Fleets}}.$$

The second ratio is the size of a given fleet to the service's total fleet size.

$$\frac{\textit{Total Number of Aircraft in the Fleet of Interest}}{\textit{Total Number of Aircraft in the Air Force}}.$$

So one might say the O&S cost of a given fleet is *X* percent of the Air Force's total aircraft O&S cost, but *Y* percent of the Air Force's primary aircraft inventory. If $X > Y$, the system is unusually costly on a per-aircraft basis, and conversely.

Cost per Capability

The commercial airline industry uses the metric *cost per available seat mile*, i.e., the cost to fly one seat one mile. This is a simple cost metric that is widely used and applicable throughout the industry. This simple and effective metric of cost and effectiveness is possible because the various fleets in the commercial industry are flown for a common and simple purpose that is easily measured.

In contrast, aircraft fleets in DoD fly a variety of missions with different purposes, many of which are complex and multidimensional. The effectiveness of DoD aircraft is often not easily measured. Therefore it is far more difficult to find cost-effectiveness measures for aircraft in DoD, and it is certainly not possible to find one cost-effectiveness metric applicable to all DoD aircraft. A cost-per-capability metric would have to be tailored and calculated for each mission area because the measure of capability differs with each mission.

Even within a given mission area, an aircraft often provides many different capabilities. Therefore, it may be impossible to estimate cost per capability across disparate aircraft when those aircraft provide different sets of capabilities. Or it may prove impossible to select a single representative capability or mission for such aircraft. There can be a variety of taskings and scenarios, so it may be impossible to fairly estimate cost per capability across disparate options when, in fact, those different options provide different sets of capabilities that are not directly comparable.

A practical complication with incorporating capability into a metric is that information on warfighting scenarios and effectiveness is often classified, which greatly restricts the use of the metric. If cost metrics are to be used as management tools, it is desirable for them to be visible and widely available to those with financial-management responsibilities.

Metrics Summary

Table 4.1 provides a summary of prospective metrics with their advantages and disadvantages.

Table 4.1
Advantages and Disadvantages of Different O&S Cost Metrics

Metric	Advantages	Disadvantages	RAND Comment
CPFH	Best for assessing costs that are variable with flying hours	Counterintuitive in the presence of fixed costs	Well suited for FHP and reimbursable rates
Cost per aircraft	Moves in same direction as total flying hours	Affected by flying hours	Preferred metric for comparing aircraft systems
Cost per squadron or unit	Could be insightful if squadrons correspond to capability	Squadron and unit sizes are not standardized	Probably not practical for cross-system comparisons
Cost per fleet	A good way to consider affordability	A different scale than CPFH or cost per aircraft	Could be challenging for cross-system comparisons
Cost per capability	Ultimately weapon systems exist to provide capability	Different weapon systems are unlikely to provide directly comparable capabilities	Probably not practical

A visual representation of the different messages conveyed by different aircraft O&S cost metrics is provided in Figure 4.2. Figure 4.2 compares KC-135 constant-dollar CPFH, cost per aircraft (PAI), and cost per fleet, normalizing each series' FY 1996 value to 1.0. (With this normalization, for each metric, Figure 4.2 displays $\frac{\textit{Year N value of metric}}{\textit{1996 value of metric}}$). Since 1996, cost per aircraft has increased considerably, but fleet costs have not increased commensurably because the Air Force retired a number of KC-135Es. Cost per flying hour declined with the beginning of increased flying hours in support of overseas operations in the early 2000s, then rose sharply as flying hours decreased as the overseas operations wound down after FY 2011.

Figure 4.2
A Comparison of KC-135 Cost per Flying Hour, Cost per Aircraft, and Cost per Fleet Since FY 1996 Normalized to FY 1996

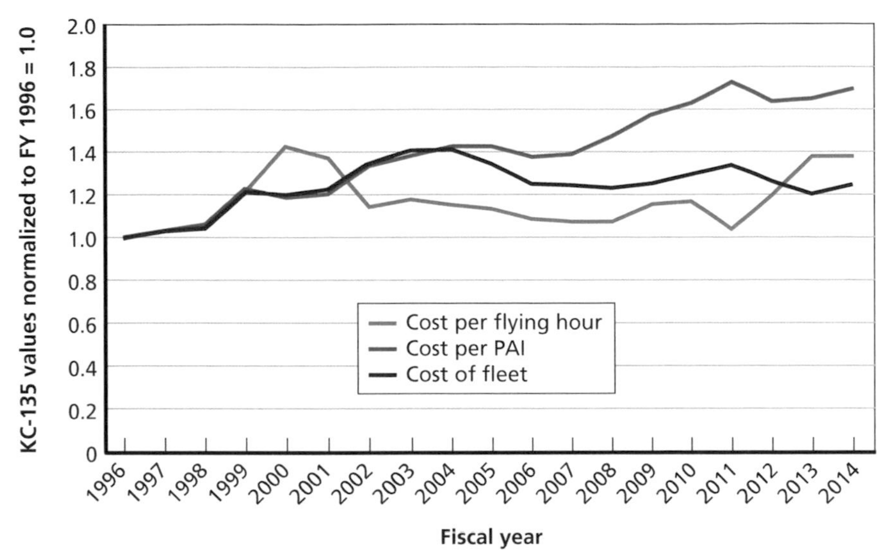

SOURCE: AFTOC query dated February 27, 2015.
RAND *RR1178-4.2*

One indication of the suitability of the cost-per-aircraft metric for comparing O&S costs of different aircraft is its use in SARs.[2] The O&S cost metrics used in 12 different aircraft program SARs are shown in Table 4.2. The table indicates that an average annual cost-per-aircraft metric is most often used by DoD to report aircraft O&S costs in SARs and compare the costs to antecedent aircraft.

---

[2] SARs are prepared annually by DoD in conjunction with submission of the President's budget and are provided to Congress. SARs contain the latest estimates of cost (including O&S cost), schedule, and performance.

Table 4.2
Different Selected Acquisition Reports' O&S Cost Metrics

Reporting Aircraft	Antecedent Aircraft	SAR Date	O&S Cost Metric
C-130J	C-130H1 & H2	Dec-13	Avg. annual cost per aircraft
E-2D AHE	E-2C	Dec-13	Avg. annual cost per aircraft
EA-18G	EA-6B	Dec-13	Avg. annual cost per aircraft
F/A-18E/F	F/A-18C/D	Dec-12	Avg. annual cost per aircraft
F-22	F-15C	Dec-10	Avg. annual cost per squadron
F-35	F-16C/D	Dec-13	Cost per flying hour
KC-46A	KC-135R&T	Dec-13	Avg. annual cost per aircraft
MQ-9 (Reaper)	MQ-1 Predator	Dec-13	Avg. annual cost per aircraft
P-8A	P-3C	Dec-13	Avg. annual cost per aircraft
RQ-4 (Global Hawk)	None	Dec-13	Avg. annual cost per aircraft
T-6 (JPATS)	T-37 (AF only)	Jun-13	Avg. annual cost per aircraft
UH-60M	UH-60L	Dec-13	Avg. annual cost per aircraft

Current OSD practices tend toward using the cost-per-aircraft metric for unitized reporting. With the exception of the F-35, SARs provide a unitized cost in the form of average annual cost per aircraft. For aircraft programs with existing affordability metrics, the goal or cap is typically set as average annual cost per aircraft, again with the exception of the F-35, which has its O&S metric set as CPFH at steady state. Of the 12 aircraft programs shown in Table 4.2, the F-22 has completed production, the program no longer submits SARs for aircraft production, and the SAR from 2010 precedes the steps taken by DoD to standardize reporting. Of the remaining 11 aircraft programs shown in the table, only the F-35 program uses a CPFH metric, and the rest use a cost-per-aircraft metric.

CHAPTER FIVE

Recommendations

In this chapter we offer recommendations on an aircraft O&S cost metric and its definition that is suitable for comparing the O&S costs of aircraft systems and similar uses in DoD. These recommendations are not intended to apply to the CPFH usages for budgeting for flying-hour programs or for calculating reimbursable rates for DoD customers.

As noted, when appropriate standardization and normalization steps are taken, CPFH can be used to compare aircraft O&S costs or to monitor trends in aircraft O&S costs over time. A CPFH metric is particularly appropriate when only the costs that are clearly variable with flying hours are of interest. The elements of cost that are included in the CPFH metric, the flying hours per aircraft, and any normalizations should be made explicit.

We and most of the subject-matter experts we interviewed view a cost-per-aircraft metric as a useful alternative metric of aircraft O&S costs. We recommend using a cost-per-aircraft metric based on the number of PAI aircraft and naming it *Direct Cost per Aircraft (PAI)* to avoid ambiguity in the definition of O&S costs and the number of aircraft.

$$\textit{Direct Cost per Aircraft (PAI)} = \frac{\textit{Direct O\&S Costs}}{\textit{PAI}}.$$

This metric has important advantages relative to CPFH for comparing aircraft O&S costs. First, assuming sizable fixed costs related to flying hours, direct cost per aircraft (PAI) will be less sensitive to changes in flying-hour-level assumptions than the CPFH metric. Second, when fixed costs are considered, as they surely will be in a cross-system O&S cost comparison, it is misleading to present these costs "per hour." Third, flying hours are relatively easy to change and contribute to the volatility of the CPFH metric, as they affect both the numerator and denominator of the CPFH metric. In contrast, the number of PAI is more stable, with large changes only when there is a restructuring of the acquisition program or existing force structure. Fourth, and perhaps most important when the metric is used as an indication of cost or affordability, direct cost per aircraft (PAI) changes in the same direction as total O&S costs. When total O&S costs for a fleet increase, the direct cost per aircraft (PAI) increases, unlike a CPFH metric that includes any fixed costs.

The direct cost per aircraft (PAI) metric should include the full spectrum of direct O&S costs associated with each weapon system, or Elements 1–5 in the DoD standard O&S cost-element structure. In some analyses, such as the comparison of U-2 and Global Hawk described in Chapter Three, it is appropriate to include costs that are not specified in the standard O&S cost-element structure, such as the cost of modifications that add additional operational capability. In these cases, it is especially important to make explicit the elements of costs that are included as well as assumptions or normalizations that affect the analysis.

Flying hours affect the cost of any aircraft O&S cost metric. To ensure comparability over time for a given aircraft or when comparing different aircraft, flying hours per aircraft per year should be specified and, ideally, normalized.

In closing, we observe that there are many possible metrics that can be used to compare the O&S costs of different aircraft. It is probably unrealistic to expect that one metric is the "best" for all purposes. As demonstrated in Figure 4.2, the metrics can convey different information about a given aircraft. By using more than one metric and focusing on those measures most relevant to the decision at hand, a fuller understanding of aircraft O&S costs can be gained.

APPENDIX A

An Example of How Cost per Flying Hour and Cost per Aircraft Change When Flying Hours Are Reduced

This appendix provides an example of how CPFH and cost-per-aircraft metrics change when flying hours are reduced. For this example, we extracted historical costs for an actual fighter aircraft and reduced all the costs by the same percentage to keep the data notional (and not For Official Use Only), yet in the same actual proportions of fixed and variable costs. Aircraft in this fighter fleet flew an average of 316 hours per aircraft in the year for which the data were obtained. Then we estimated the annual costs based on an average 250 flying hours per aircraft per year, keeping the costs the same in both cases for the elements denoted as "fixed" in the right-most column of Table A.1 and multiplying the costs denoted as "variable" in the right-most column by 250/316. The latter adjustment for variable costs reduces those costs in direct proportion to the reduction in flying hours.

This example assumes 192 PAI under either flying-hour scenario.

We chose the aircraft for this example because it had nearly the same FH per PAI per year that the Air Force had originally planned to fly the F-35A. (The Air Force originally planned to fly its F-35s 300 FH per PAI per year.) Reducing the flying hours of this example to 250 FH per PAI per year would be approximately the same percentage change as what occurred when the Air Force reduced the F-35A flying hours from 300 to 250 FH per PAI.

Table A.1 displays the data used in this example and the calculations we made.

Table A.1
An Illustration of the Effect of Reduced Flying Hours on Cost per Aircraft and Cost per Flying Hour Using Notional Data

O&S Cost Element	Annual O&S Cost $M, 316 FH/PAI	Annual O&S Cost $M, 250 FH/PAI	RAND Assessment of Relationship to Flying Hours
1.0 Unit personnel	385	385	Fixed
2.0 Unit operations	450	356	Variable
3.1 Consumable materials/repair parts	68	54	Variable
3.2 Depot-level reparables	349	276	Variable
3.4 Engine depot maintenance	7	6	Variable
3.4 Depot maintenance (other than engine)	107	107	Fixed
4.0 Other sustaining support	12	12	Fixed
5.0 Modifications	83	83	Fixed
Total annual O&S, $M	1,461	1,279	
Direct cost per aircraft (PAI), $M	7.6	6.7	
CPFH, $K	24.0	26.6	

The leftmost column of Table A.1 shows the cost-element number of the DoD CAPE's O&S cost-element structure and its nomenclature.

The second column from the left has the cost data at the reported flying-hour rate of 316 FH per PAI. Before normalization for flying hours, the raw data total $1.461 billion for the annual cost of this fleet of aircraft.

The third column from the left has the annual costs adjusted to 250 FH per PAI. The elements of petroleum, oil and lubricants; consumable materials/repair parts; depot-level reparables; engine-depot maintenance; training munitions and expendables; and other unit-level consumption are variable with flying hours. As expected with fewer total flying hours consuming fewer resources, the total annual cost for this fleet of aircraft drops to $1.279 billion.

The rightmost column denotes the elements that we assumed to be fixed and variable with flying hours. Costs that are fixed are the same in the two middle columns regardless of the reduction in flying hours. Costs that are variable are reduced by 250/316 from the second column to the third column in proportion with the reduction in flying hours.

The bottom two rows of Table A.1 show the effects on two cost metrics: cost per aircraft and CPFH. The reduction of 12 percent in total annual fleet costs is mirrored in a decrease in cost per aircraft of 12 percent, or $0.9 million. Counterintuitively, as a measure of affordability, CPFH increases roughly $2,600, or 11 percent. The results are shown in percentage terms in Figure A.1.

Figure A.1
Effects of a Reduction in Flying Hours on Cost per Flying Hour and Cost per Aircraft

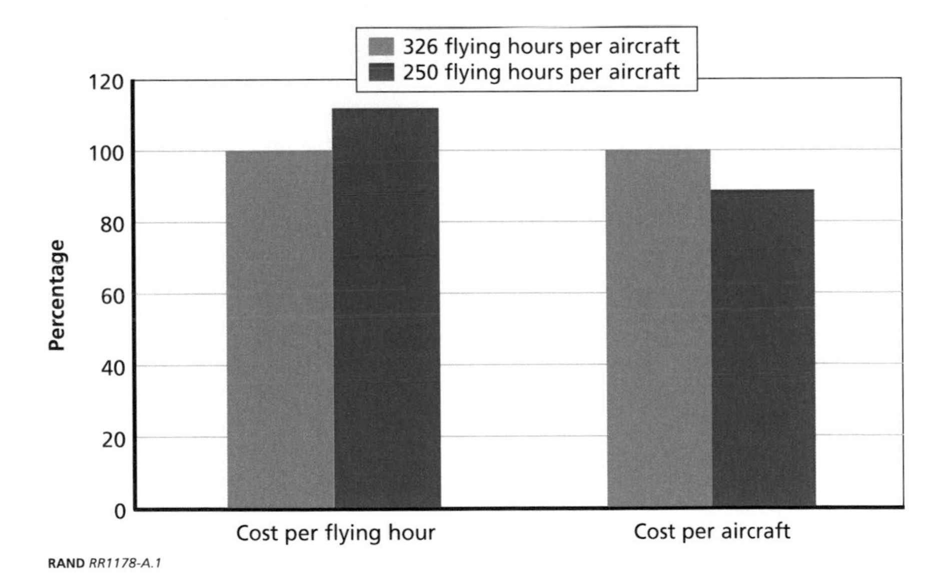

APPENDIX B

Statistical Analysis of the Relationship Between Aircraft Annual O&S Costs and Number of Aircraft and Flying Hours

Introduction

The analytical approach contained in this appendix is based primarily on the work of Unger (2009), in which the author examined the relationship between the annual inventory and flying hours by mission/design (MD) fleet and annual O&S costs of U.S. Air Force aircraft. Unger's analysis used annual data from FY 1996 through 2006 at the MD level. For total yearly O&S costs (that is, all the elements of the standard DoD O&S cost-element structure), Unger found that costs increased roughly 6 percent with a 10-percent increase in flying hours. Results at lower levels of the cost-element structure for some elements of interest including depot-level reparables and depot maintenance costs were inconclusive. An inherent and difficult problem in the statistical analysis is the very high correlation between aircraft inventory and flying hours, a phenomenon in regression analysis called *multicollinearity*, which makes problematic the interpretation of the coefficient estimates of highly correlated independent variables. Since the publication of Unger's report, the DoD standard cost-element structure changed, and there are also several years of additional cost and flying hour data available. In addition, in the hopes of finding greater variation between aircraft inventory and flying hours in the data, we performed an analysis of the data at the major command (CMD) level rather than the total fleet level for each aircraft MD. The following presents the results of a statistical analysis of the relationship between O&S cost, aircraft inventory, and usage using a similar statistical approach to Unger's, but that accounts for the new cost structure, incorporates the additional years of data, and extends the analysis of the data to an additional level of detail at the command level.

Data Overview and Estimation Approach

The analysis is based on AFTOC data for FY 1996 through 2011. AFTOC tabulates costs by FY, cost element, and weapon system, and provides costs in either nominal or constant dollars. We used constant-dollar data by aircraft MD and command level. We wanted to analyze only those commands for which an MD had a substantial number of flying hours in a given year. Therefore, any command for which FH for a given MD in a year was less than 10 percent of the total FH for that MD in that year was excluded from the analysis.

One approach to assess this relationship would be an MD-level linear regression, e.g., regress B-1 O&S costs on B-1 flying hours and B-1 TAI. A weakness of this approach is a small sample size—i.e., 16 years of data at the total B-1 fleet level. In order to assess a more general relationship between costs and flying hours and fleet size that does not suffer from a small sample size, we chose to estimate an ordinary least-squares regression of the form

$$ln(Cost_{ijt}) = a_{ij} + b * ln(FN_{ijt}) + c * ln(TAI_{ijt}) + d * Year_t + e * MD_i + f * CMD_j + \varepsilon_{ijt},$$

where each i is an MD, each j is a CMD, and each t is a year. The *Year*, *MD*, and *CMD* terms are dummy variables, which each has its own intercept; d, e, and f, estimated by the regression, respectively. The coefficient a is the intercept for each MD and CMD combination. In this estimation each MD-CMD combination gets its own intercept, but there is a common b that is the relationship between the natural log of flying hours and the natural log of costs and a common *c* that is the relationship between the natural log of TAI and the natural log of costs. Intuitively, *b* is the estimated elasticity of cost with respect to flying hours. If flying hours increase 1 percent, on average, costs increase *b* percent. If $b > 1$, costs increase disproportionately as flying hours increase. If $b = 1$, costs grow in proportion to flying hours. If $b < 1$, costs do not increase in proportion to flying hours. The same logic applies to *c* for the relationship between cost and TAI.

Estimation Results

In this section, we present a series of regressions with the natural log of various cost elements as the dependent variables and the natural log of flying hours, the natural log of TAI, and the FY, MD, and CMD dummy variables as the independent variables.

First, we present the highest-level regression with the natural log of total costs of all O&S cost elements as the dependent variable. Table B.1 shows the result. The key result is its Ln(FH) coefficient estimate of 0.43392 and the Ln(TAI) coefficient of 0.36514. This result suggests both flying hours and fleet size have partial effects on total costs, with both coefficient estimates being significantly greater than zero but also significantly less than one. The Pearson correlation between Ln(FH) and Ln(TAI) in our data is 0.9333, indicating a very high amount of multicollinearity and suggesting that the magnitude of the coefficients on those variables should be treated with caution.

The FY coefficients are measured relative to the omitted year, 1996. The MD coefficients are measured relative to the omitted MD, the C-130. The CMD coefficients are measured relative to the omitted CMD, Air Combat Command. It is important for analytic purposes that these independent variables are included, but we do not think that they have great policy importance. For this reason, we list the MDs and CMDs in the following two sentences, but not in the regression results. MDs included in the regression are A-10, AC-130, AT-38, B-1, B-2, B-52, C-141, C-17, C-20, C-21, C-26, C-37, C-5, C-9, E-3, E-8, EC-130, F-117, F-15, F-16, F-22, HC-130, KC-10, KC-135, LC-130,

MC-130, RC-135, T-1, T-37, T-38, T-43, T-6, U-2, and WC-130. Commands included in the regression are AETC, AFGSC, AFMC, AFRC, AFSOC, AMC, ANG, PACAF, and USAFE.

Table B.1
Total Spending Regression with Flying Hours and Fleet Size as Independent Variables

Observations	967
F(60,906)	424.4
Prob > F	0.0000
R-squared	0.9634
Dependent variable	Ln (total spending all CAPE categories)

Independent Variable	Coefficient	SE	T	P >|t|
Ln(FH)	0.43392	0.03375	12.859	< 2e-16
Ln(TAI)	0.36514	0.03771	9.68	< 2e-16
Constant	14.00230	0.24711	56.67	< 2e-16

In order to examine the relationship of more detailed cost elements and flying hours and fleet size, we ran several log-log regressions, with the natural log of various cost elements as the dependent variable. Table B.2 shows the result. The left side of the table shows the dependent variable, the middle of the table shows the coefficient of Ln(FH), and the right side of the table shows the coefficient of Ln(TAI). It should be noted that not all of these regressions cover as many aircraft as Table B.1, since some of these cost categories had no expenditures for some aircraft. In addition, we excluded observations for some aircraft that were heavily supported by contractor logistics support, because AFTOC is unable to allocate such costs to the appropriate detailed cost elements. These aircraft did not present a problem for analysis of total O&S costs.

Table B.2's coefficients are difficult to interpret by themselves. In order to show the implications of the results, Table B.3 puts the coefficients into categories that indicate the nature of the relationship between each element of cost and aircraft inventory or FH. The logic for deriving Table B.3 is as follows. There is no relationship between a variable and cost if we can reject b (or c) = 1 but cannot reject that b (or c) = 0 based on the data shown in Table B.2. There is a partial relationship if we can reject b (or c) = 1 and we can reject b (or c) = 0. Finally, there is a proportional relationship if we cannot reject that b (or c) = 1 but we can reject b (or c) = 0. Table B.2 shows that Elements 3.5, Other Maintenance, and 5.0, Continuing System Improvements, are incoherent, and there were no data for 3.3, Intermediate Maintenance, so these elements will not appear in Table B.3.

Table B.2
Level-One Ln(FH) and Ln(TAI) Regression Coefficient Estimates

		Ln(FH) Coefficient		Ln(TAI) Coefficient	
Dependent Variable	**Element Nomenclature**	**Estimate**	**SE**	**Estimate**	**SE**
Ln(total spending)	Total spending	0.43392	0.03375	0.36514	0.03771
Ln(CAPE 1.0)	Unit personnel	0.42699	0.04337	0.28697	0.04296
Ln(CAPE 1.1)	Operations personnel	0.53346	0.19758	0.20160	0.19564
Ln(CAPE 1.2)	Maintenance personnel	0.23122	0.07959	0.47336	0.07891
Ln(CAPE 1.3)	Other direct support personnel	0.65391	0.07291	0.23535	0.07222
Ln(CAPE 2.0)	Unit operations	0.90994	0.03508	0.05692	0.03474
Ln(CAPE 2.1)	Operating material	0.97854	0.05711	0.01130	0.05655
Ln(CAPE 2.2)	Support services	0.65292	0.15581	0.47285	0.15453
Ln(CAPE 2.3)	Temporary Duty	0.17607	0.10762	0.32136	0.10693
Ln(CAPE 2.4)	Transportation	–0.60622	0.29927	1.02736	0.29761
Ln(CAPE 3.0)	Maintenance	0.31338	0.09516	0.57945	0.09464
Ln(CAPE 3.1)	Consumable materials and repair parts	0.48328	0.06279	0.42721	0.06236
Ln(CAPE 3.2)	Depot-level reparables	0.36764	0.07679	0.56812	0.07626

Table B.2—Continued

		Ln(FH) Coefficient		Ln(TAI) Coefficient	
Dependent Variable	**Element Nomenclature**	**Estimate**	**SE**	**Estimate**	**SE**
Ln(CAPE 3.3)	Intermediate maintenance	No Data			
Ln(CAPE 3.4)	Depot maintenance	0.01791	0.13606	1.03204	0.14124
Ln(CAPE 3.5)	Other maintenance	–1.70695	0.79568	1.86856	0.94008
Ln(CAPE 4.0)	Sustaining support	0.23414	0.23552	0.20721	0.23364
Ln(CAPE 5.0)	Continuing system improvements	–0.05026	0.49858	0.54859	0.59538
Ln(CAPE 6.0)	Indirect support	0.03282	0.10595	0.91870	0.10495

Table B.3
Summary of Ln(FH) and Ln(TAI) Regression Coefficient Estimates

		Ln(FH) Coefficient		
		No Relationship to FH	**Partial Relationship to FH**	**Proportional to FH**
	No relationship to TAI	4.0 Sustaining support	1.1 Operations personnel	2.1 Operating material
Ln (TAI) coefficient	**Partial relationship to TAI**	2.3 TDY	Total spending 1.2 Maintenance personnel 1.3 Other direct support personnel 2.2 Support services 3.1 Consumable material and repair parts 3.2 DLRs	
	Proportional to TAI	2.4 Transportation 3.4 Depot maintenance 6.0 Indirect support		

In total, these results provide some basis for the RAND-assessed relationship between cost category and flying hours presented in Table 2.1 of this report. Table B.4 repeats the information from Table 2.1 and includes the statistical relationship between Ln(cost) and Ln(FH) as a result of this analysis.

Given the high correlation between flying hours and fleet size, or multicollinearity between these independent variables, one must be cautious when interpreting the statistical relationship between flying hours and cost. The statistical analysis was consistent with the RAND assessments in many instances, including Elements 2.1, Operating Material; 2.3, TDY; 2.4, Transportation; 3.4, Depot Maintenance; 4.0, Sustaining Support; and 6.0, Indirect Support. The data provided incoherent results for 3.4, Depot Maintenance, and 5.0, Continuing

Table B.4
Statistical Evaluation of RAND-Assessed Relationship Between Cost and Flying Hours

Element Number and Nomenclature	RAND-Assessed Relationship to Flying Hours	Statistical Relationship to Flying Hours
1.0 Unit-Level Manpower	Fixed	Partial
1.1 Operations	Fixed	Partial
1.2 Unit-level maintenance	Fixed	Partial
1.3 Other unit level	Fixed	Partial
2.0 Unit Operations	—	—
2.1 Operating material	Variable	Proportional
2.2 Support services	Fixed	Partial
2.3 Temporary duty	Fixed	None
2.4 Transportation	Fixed	None
3.0 Maintenance	—	—
3.1 Consumable materials and repair parts	Variable	Partial
3.2 Depot-level reparables	Variable	Partial
3.3 Intermediate maintenance	Variable	No data
3.4 Depot maintenance	Semivariable	None

Table B.4—Continued

Element Number and Nomenclature	RAND-Assessed Relationship to Flying Hours	Statistic Relationship to Flying Hours
3.5 Other maintenance	Undefined-Unknown	Incoherent
4.0 Sustaining Support	Fixed	None
4.1 System-specific training	Fixed	—
4.2 Support equipment replacement and repair	Fixed	—
4.3 Sustaining/systems engineering	Fixed	—
4.4 Program management	Fixed	—
4.5 Information systems	Fixed	—
4.6 Data and technical publications	Fixed	—
4.7 Simulator operations and repair	Fixed	—
4.8 Other sustaining support	Fixed	—
5.0 Continuing System Improvements	Fixed	Incoherent
5.1 Hardware modifications	Fixed	—
5.2 Software maintenance	Fixed	—
6.0 Indirect Support	Fixed	None
6.1 Installation support	Fixed	—
6.2 Personnel support	Fixed	—
6.3 General training and education	Fixed	—

System Improvements, so no conclusions can be made regarding those elements. For the remaining elements, the statistical analysis showed that there was a partial relationship between flying hours and cost. In most of these cases, it was difficult to distinguish between the variability due to flying hours versus fleet size.

In an effort to investigate some of these elements further, we performed the analysis at yet another level of detail for 3.1, Consumable Materials and Repair Parts, and 3.2, Depot-Level Reparables. These two elements were of particular interest because they are widely considered to be highly variable with FH, as indicated by their inclusion in the costs budgeted in the FHP. Rather than aggregating to the command level as before, we aggregated flying hours, TAI, and costs at the base level. We hoped that variation between flying hours and TAI would increase at the base level compared to the command level, and would therefore decrease the correlation between these independent variables and result in easier to interpret coefficient estimates. The correlation between Ln(FH) and Ln(TAI) at the base level was 0.921, not a significant reduction from the correlation at the command level. The resulting coefficients are shown in Table B.5.

Table B.5
Summary of Ln(FH) and Ln(TAI) Coefficient Estimates Using Base-Level Data

		Ln(FH) Coefficient		Ln(TAI) Coefficient	
Dependent Variable	Element Nomenclature	Estimate	SE	Estimate	SE
Ln(CAPE 3.1)	Consumable materials and repair parts	0.24292	0.09083	0.70131	0.12012
Ln(CAPE 3.2)	Depot-level reparables	0.12971	0.13366	1.21941	0.17622

Element 3.1, Consumable Materials and Repair Parts, remained partially variable with both FH and TAI, but with a much higher coefficient on TAI than in the command-level analysis. Element 3.2, Depot Level Reparables, showed no statistically significant relationship to FH

and a greater-than-proportional relationship to TAI in the base-level analysis, a big change from the results of the command-level analysis. Both statistical results for depot-level reparables are counterintuitive, but the base-level results showing no relationship to FH are especially puzzling. The significant changes in the magnitude and significance of the coefficients on TAI and FH when the aggregation of data changes from the command level to the base level, coupled with the very high multicollinearity in both aggregations of data, suggest that the results need to be considered with a great deal of caution.

References

Air Force Total Ownership Cost (AFTOC) decision support system, available with restricted access. As of November 1, 2014:
https://aftoc.hill.af.mil

Anneker, Monique, Daniel Germony, and Todd Pardoe, *Develop a Better Aircraft Cost Metric*, thesis, Monterey, Calif.: Naval Postgraduate School, 2014.

Chairman of the Joint Chiefs of Staff, *Standardized Terminology for Aircraft Inventory Management*, Chairman of the Joint Chiefs of Staff Instruction (CJCSI) 4410.01G, October 11, 2013.

CNN.com, "Official Who OK'd Air Force One Jet Flyover Resigns," *cnn.com*, May 8, 2009, 8:59 p.m. EST. As of October 29, 2014:
http://www.cnn.com/2009/POLITICS/05/08/air.force.one.flyover/index.html

Gates, Robert M., letter to Senator John McCain, Washington, D.C., May 5, 2009.

Office of the Secretary of Defense, *Cost Assessment and Program Evaluation, Operating and Support Cost-Estimating Guide*, March 2014. As of April 27, 2015:
http://www.cape.osd.mil/files/OS_Guide_v9_March_2014.pdf

Office of the Under Secretary of Defense (Comptroller), *Fiscal Year (FY) 2014 Department of Defense (DoD) Fixed Wing and Helicopter Reimbursement Rates*, Washington, D.C., Department of Defense, October 1, 2013. As of September 5, 2014:
http://comptroller.defense.gov/Portals/45/documents/rates/fy2014/2014_f_h.pdf

Robbert, Albert A., *Costs of Flying Units in Air Force Active and Reserve Components*, Santa Monica, Calif.: RAND Corporation, TR-1275-AF, 2013. As of April 27, 2015:
http://www.rand.org/pubs/technical_reports/TR1275.html

Under Secretary of Defense for Acquisition, Technology, and Logistics, *Operational Support Airlift (OSA)*, Department of Defense Instruction 4500.43, Incorporating Change 1, June 26, 2013. As of April 27, 2015:
http://www.dtic.mil/whs/directives/corres/pdf/450043p.pdf

Unger, Eric J., *An Examination of the Relationship Between Usage and Operating-and-Support Costs of U.S. Air Force Aircraft*, Santa Monica, Calif.: RAND Corporation, TR-594-AF, 2009. As of April 27, 2015:
http://www.rand.org/pubs/technical_reports/TR594.html

U.S. Air Force, *Global Hawk Block 30 Divestiture*, Washington, D.C., April 2013.

———, *United States Air Force Working Capital Fund (Appropriation: 4930): Fiscal Year (FY) 2015 Budget Estimates*, March 2014. As of February 28, 2015: http://www.saffm.hq.af.mil/shared/media/document/AFD-140310-016.pdf

U.S. Department of Defense, "Defense Working Capital Funds Activity Group Analysis," *DoD Financial Management Regulation, DoD 7000.14-R*, Volume 2B, Chapter Nine, November 2009. As of April 27, 2015:
http://comptroller.defense.gov/Portals/45/documents/fmr/archive/02barch/02b_09_Nov09.pdf

———, "Operation and Maintenance Appropriations," *DoD Financial Management Regulation, DoD 7000.14-R*, Volume 2A, Chapter 3, September 2010. As of August 24, 2015:
http://comptroller.defense.gov/Portals/45/documents/fmr/Volume_02a.pdf

———, "Collections for Reimbursements of DoD-Owned Aircraft (Fixed Wing)," *DoD Financial Management Regulation, DoD 7000.14-R*, Volume 11A, Chapter Six, Appendix E, January 2011. As of April 27, 2015:
http://comptroller.defense.gov/Portals/45/documents/fmr/Volume_11a.pdf

U.S. Government Accountability Office, *Defense Acquisitions: Assessments Needed to Address V-22 Aircraft Operational and Cost Concerns to Define Future Investments*, Washington, D.C., GAO-09-482, May 11, 2009. As of April 27, 2015:
http://www.gao.gov/products/GAO-09-482

———, *Defense Logistics: Improvements Needed to Enhance Oversight of Estimated Long-term Costs for Operating and Supporting Major Weapon Systems*, Washington, D.C.: GAO-12-340, February 2, 2012. As of April 27, 2015:
http://www.gao.gov/products/GAO-12-340

———, *F-35 Sustainment: Need for Affordable Strategy, Greater Attention to Risks, and Improved Cost Estimates*, Washington, D.C.: GAO-14-778, September 23, 2014. As of April 27, 2015:
http://www.gao.gov/products/GAO-14-778

Wallace, John M., Scott A. Houser, and David A. Lee, *A Physics-Based Alternative to Cost-Per-Flying-Hour Models of Aircraft Consumption Costs*, McLean, Va.: Logistics Management Institute, Report AF909T1, August 2000.

World Maritime News, "US Costs for MH 370 Search Reach $11.4 Mln," April 25, 2014. As of August 24, 2015:
http://worldmaritimenews.com/archives/122043/us-costs-for-mh-370-search-reach-usd-11-4-mln